Pre cast and Pre stress Technolog y

2014

The Precast technology dates as far back as the Roman civilization, where it was commonly employed in building complex network of aqueducts, culverts and tunnels. The production of precast concrete in a controlled environment enables engineers to properly cure the concrete. It also avails the employees, the opportunity to closely monitor the curing process in the precast plant. Today, Precast Concrete system is being used extensively, in a wide array of construction projects. The major aim of using this construction technique is to improve the strength as well as the performance of the structures. This technique is being applied in the construction of wall panels, I-beam girders, concrete beam, spandrels, columns, spandrels, single and double tees, segmental bridge units, bulb-tee girders etc. The construction materials that are manufactured through this way are usually crack-free under working loads. Thus, prestressed materials are more watertight and also have better looks. Prestressed technology is one aspect of construction technology, that is been widely used in development of a wide array of structures, especially bridge structures. During this occasion, this book is created aiming narrowed to the student to help their understanding in term of benefits, method, application and also expandable for future technology.

Process, Method and Future Technology

Table of Contents

CHAPTER 1

INTRODUCTION

1.1. WHAT IS PRECAST

In construction industry, the term "Precast" is used to describe a construction product, which is manufactured by casting concrete in a reusable mold or "form". The distinctive reusable mold is then cured in a controlled environment, after which it's transferred to site, where it's eventually lifted into place [1]. This construction methodology is quite different from the standard method, where concrete mixtures are poured into site-specific forms and then cured on site. Precast concrete should not be confused with the precast stone. Even though both of them have the appearance of naturally occurring rock or stone, precast concrete can be distinguished by using a fine aggregate in the mixture [1].

The production of precast concrete in a controlled environment enables engineers to properly cure the concrete. It also avails the employees, the opportunity to closely monitor the curing process in the precast plant. Today, Precast Concrete system is being used extensively, in a wide array of construction projects. The wide usage of this technology has been attributed to its many advantages over the standard concrete that are usually produced on-site. For instance, the production process for Precast Concrete is usually carried out on ground level. This helps to increase the safety of workers, throughout the course of production. Also, the engineers have greater control of the quality of materials as well as the workmanship in a precast plant. The production process is also cheaper as the forms used in a precast plant, can be reused several times before being replaced [2].

The Precast technology dates as far back as the Roman civilization, where it was commonly employed in building complex network of aqueducts, culverts and tunnels. The use of precast technology in the modern world started in Liverpool, England, 1905. The idea was constructed by two engineers John Alexander Brodie and Yannick Macken. Even though the concept wasn't adopted extensively in Britain, it nevertheless expanded to all over the world, especially in Eastern Europe and Scandinavia [2].

Today, precast technology is one of the most commonly used technologies in the construction and building industries [3]. It's widely used in lots of architectural and structural applications, which can feature either parts of or an entire building system. Currently, the precast technology is focusing mainly on those concrete elements that can be used in the building of above-ground structures. Some good examples include: buildings, parking structures and bridges.

1.2. BENEFITS OF PRECAST

Building and constructed infrastructures are naturally exposed to everyday wear and tear. Unfortunately, this exposure tends to speed up the dilapidation rate of these structures. This has made it very essential to make use of construction materials that are can withstand the effects of this inevitable exposure. This is where the use of precast concrete really makes sense. Basically, precast concretes are hard and the surfaces are so tough and able to withstand the everyday wear and tear [3]. Precast concrete, a highly efficient, practical method of concrete construction makes beautiful buildings possible at a cost that rivals even the most utilitarian industrial building. Some of the benefits of precast technology are listed as follows:

Design-Build Efficiency: Using precast concrete is the surest way of obtaining an efficient delivery model for any building and construction project. For instance, the use of precast concrete enables both the building construction and the design to proceed simultaneously. This help to boost the efficiency of the design-building process of the project [3].

Resistant to Elements: As already stated earlier, precast concrete are highly resistant to elements that bring about wear and tear, oxidation etc. There is no limit to this resisting capability. Precast concrete structures have superb ability to withstand natural disasters, fires, insects, friction, oxidation, mold, mildew etc. As a result of this, the maintenance and insurance cost of precast concrete structures are relatively lower than what is obtainable elsewhere [3].

Energy Efficiency: Precast concrete structures possess inherent thermal mass that help to improve energy efficiency and cut down the heating and cooling peaks and loads of the production process. This feature often necessitates less costly mechanical systems [3].

Environmentally Friendly: The natural ability to resist mold, is one of the most important inherent characteristics of precast concrete. This minimizes the off gassing and the production of Volatile Organic Compounds (VOCs), thereby reducing the health concerns from these compounds. Precast concrete structures are also 100% recyclable. With these, precast concrete technology meets the ever increasing demand for sustainable construction and design, which are environmentally friendly [3].

Highly Esthetics: Precast concrete are highly esthetics. Thus, it's always possible to add limitless collection of textures, colorings, and patterns to the concrete mix. It's also possible to use any textured paints of any color to produce specific exceptional effects. Similarly, precast structure can be formed in any size, shape and texture. This functionality has made it possible to produce designs that can either deflect or absorb sound. This explains why precast structures are widely used as good acoustic materials for music and effective sound barriers along busy roads [3].

Fastness: Generally, it's always faster to execute a construction project with total precast system. This has the ability to save the engineers between six and eight week compared to other methods. Thus, precast technology remains the best option to use, when working under a tight schedule. It can also be critical in bringing a new building into a competitive market [3].

Maintenance: Generally, the maintenance costs for precast structures are relatively lower and affordable. In most cases, painting the surfaces of such precast concrete structures aren't mandatory, as the structures are very resistant to elements and can therefore be left unpainted without causing any damage. In cases where painting is needed, the structure can stay as long as five to ten years, before repainting is required. Finally, the interior of precast concrete structures are easier to wash and less susceptible to damage [3].

Sound Control: Precast technology is the right choice for designing buildings with acoustics properties. Unlike wooded and steel structures, precast concrete structures have the added advantage of absorbing sound. Thus, it's the right material for achieving sound control [3].

1.3. PRE-STRESS TECHNOLOGY AS A METHOD OF PRECASTING

The term "Prestressing" refers to a special construction technique, which is used to introduce stresses into a structural material, during its construction [4]. The major aim of using this construction technique is to improve the strength as well as the performance of the structures. This technique is being applied in the construction of wall panels, I-beam girders, concrete beam, spandrels, columns, spandrels, single and double tees, segmental bridge units, bulb-tee girders etc. The construction materials that are manufactured through this way are usually crack-free under working loads. Thus, prestressed materials are more watertight and also have better looks. Such materials provide a better corrosion protection for the steel materials. It has been discovered that the costs of producing prestressed materials are relatively lower. Prestressed structures also have lifetime maintenance.

Prestressed technology is one aspect of construction technology, that is been widely used in development of a wide array of structures, especially bridge structures. The technology is synonymous with precast concrete and has therefore being recognized as a method of pecasting technology. Over the years, a number of innovative technologies have been developed by several scientists, to boost the structural performance of Prestressed concrete structures as well as their long-term durability [5] These include the development of novel structural systems and the advancement in construction materials.

CHAPTER 2

PROCESS OF INSTALLATION MEMBER OF PRECAST

Figure 1 : A typical flow of the production process of precast concrete components

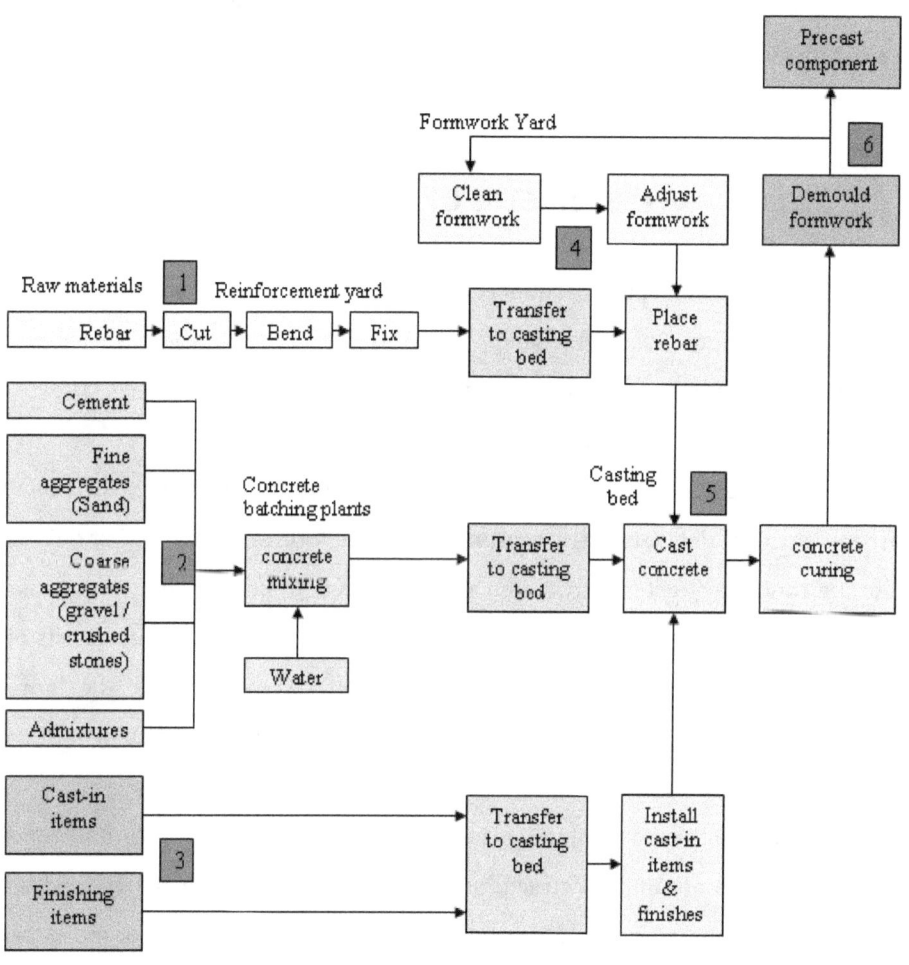

Source: CEBE Working Paper (2013) [6]

Generally, the production process of precast structures consists of several phases. According to the working paper of JIT [6], these phases can be grouped into six different groups namely;

- cutting, bending and fixing reinforcement bars;

- mixing concrete in in-house concrete batching plants;
- preparing cast-in items (e.g prestressing tendons) and finishes requirements;
- moulding formwork;
- casting and curing the concrete components; and
- demoulding formwork to get the finished precast components.

The second phase can be overlooked, if an external ready-mixed concrete is being used in the plant. In this section, we discussed the most essential phases of the production process, which are: process of formwork, concreting, casting, curing, and transport to site.

2.1. PROCESS OF FORMWORK

The two factors that have the utmost impact on the operation of every precast plant are the formwork and concrete [7] High quality formwork systems are very instrumental in obtaining a quality, smooth precast concrete finish.

Formworks are frameworks, used to offer temporary support for a permanent structure that is under construction, until it is self-supporting [7]. It's the formwork system that defines the shape of the concrete structures that is being constructed. It includes the surface, support and framing, which are used to accomplish this objective. In other words, it consists of the forms on which the concretes are poured, the support system that hold the form and concrete as well as the foundations and footings. At times, bracings may also be included to boost the stability of the structures.

For formwork to fulfill the main objectives for its development, there are certain properties that must be incorporate into its structural designs. These include: rigidness and water-tightness (8). During construction activities, the proper position and shape of the formwork should be maintained by tying and bracing the components together. The formwork systems should also be constructed in a way that makes it possible to be removed easily and safely, without damaging the formed concrete. If the formwork is going to be reused, then design should be done in such a way, as to ensure the retention of the form strength after allowing for the deterioration of materials through use and handling [8].

There are several types of formwork systems that are currently being used in the construction industry. Some of these examples include: steel plate forms, tilting tables, battery moulds, carrousel systems with production pallets etc. The exact type to be used in any case is normally determined by the volume of production of a particular element and flexibility desired in production [8]

The materials that can be used in the construction of precast forms include: Wood, Steel, Aluminium, Fiberglass, Plastic, Concrete, Expanded Polystyrene. The ability of the material to retain its shape against the hydrostatic pressure of the concrete, withstand the vibrations, provides product tolerance are the major factors to consider, when choosing the materials to be used in constructing the formwork.

Generally, wall panels are either cast separately or unto a long bed when prestressed. The bottom of the frameworks that are normally used in the production of wall panels consists of steel plate, with a minimum thickness of about 5mm. The sides of the formwork are typically made up of fixed rail/channel or wooden materials, while the blockouts are manufactured with wooden materials [8]. Two common examples of formwork that can be used for precast wall panels are Tilting tables and Battery Mould.

Figure 2: Battery Mould

Source: Hardas (2013) [8]

Figure 3: Tilting Table

Source: Hardas (2013) [8]

There are also specific formwork for constructing Columns, beams and hollowcore slabs. These formworks can be constructed with steel or wood.

2.2. CONCRETING AND CASTING

Concreting is another important phase of precast construction. The key to a successful precast elements, include good mix design and careful placing [9]. These two strategies help to prevent unwanted air bubbles, which can be detrimental to the quality of the precast concrete structures. Using poor quality concrete can also affect the durability performance of the concrete structures, negatively. Thus, it's very necessary to always monitor the concrete quality and cover.

Generally, precast elements are manufactured in the factory. The concrete mixes that are commonly used consist of; cement, aggregates from quarries, and other special ingredients. The mixed concrete is then introduced into a form, where it's often subjected to prestressing and curing. Once the elements are cured, the final products are then stripped from the form and transported to the precasters yard for finishing and storage prior to shipping to the jobsite.

In today's contemporary society, there has been increasing needs to include special chemicals during the concreting process. The essence of this is to improve the cycle times, clean and maintain formwork, and boost the quality of the final products. A typical precast concrete mix consists of about 10% to 12% cement by volume. The main function of the cement is to react with water, and bind all the ingredients of the mix.

Studies have shown that the production process of cement, contributes about 1% to 2% of global carbon dioxide emissions [10]. This is mainly generated through the burning of fossil fuels and process-related emissions. Thus, the production process of concrete elements can be made more eco-friendly by minimizing the amount of cement to be used. The amount of cement used in precast concrete may be reduced by up to 60% through substitution by supplementary cementitious materials (SCMs). However, the possibility of cement substitution are determined by several factors such as mixture design requirements, the products and processes of individual precast concrete manufacturers and plants, and the local availability of materials [10].

During the casting process, almost all the faces are shuttered in formwork. The only exception is the top side, which must remain exposed. It's actually through this face that the concrete is poured and then wood-floated or steel-floated. Thus, the top-face will always look different from the other sides of the structures. Because of this, the top face to choose should be determined by the structure that is being produced. The choices are as follows [10]:

- For precast beams, it is the invisible side that will have the deck slab cast on it.
- For precast noise barriers, it is the bottom or back face.
- For precast sidewalk panels, it is the bottom surface.
- For precast parapets it is the narrow top face

2.3. CURING

Curing is the hydration reaction that is responsible for hardening of the concrete. The reaction normally commenced, after mixing water with cement and the aggregates that make up the concrete. The process can be split into three major stages. These are as follows [11]:

Stage 1: This stage commences immediately water has been added to the cement and dry concrete mix, and then last till the onset of the initial set, whose compressive strength has been designated 500 psi by some engineers [1]. It's during this stage that the reaction between cement and water begins. The measurable compressive strength that is acquired during this stage is minimal. The stage 1 of the curing process can last for 3-4 hours. The exact duration is normally determined by the materials, being used in construction of the structures [6].

Stage 2: The stage 2 of the curing process begins at the onset of initial set. During this stage, the rate of hydration is usually increased and a consequent release of exothermic heat. There is also a rapid compressive strength gain. The overall duration of this second stage of the curing process, is hugely determined by the particular mix design in use and the curing conditions. For common precast applications, the stage two of the curing process can last between six to eight hours. Also, the rate of compressive strength development can vary between 500 to 700 psi per hour [12].

Stage 3: This stage begins, after most of the cementitious materials must have reacted with water. The intensity of heat released at this stage, is significantly lower than the previous stage. The rate of strength development reduced to between 50 to 100 psi per hour. Generally, the rate of strength gain is rarely affected by elevating the curing temperature [12].

There are two basic methods that can be used to manipulate the hydration process and obtain high early compressive strength in concrete structures. First, the environmental curing conditions can be altered to increase the curing temperature and bring about an elevated rate of the hydration reaction [4]. Secondly, the structural engineers on duty can maximize the initial rate of compressive strength gain, by specifying the cement composition [7]. This can be achieved by choosing the right cement type. The initial rate of compressive strength gain can also be maximized by using both mineral and chemical admixtures. The most economical accelerated curing process can be accomplished by combing the various available methods.

A lot of researches have been conducted to determine the relationship between curing temperature and the rate of compressive strength gain in concrete structures. This has led to the establishment of rules guiding the rate of compressive strength gain. For instance, it's now known that the increase in curing temperature is directly proportional to the rate of strength gain [9] Thus, once the curing temperature is increased, a complementary increase in the rate of strength gain will also be recorded. However, this proportionality is only valid till a certain critical point, when the increases in temperature does not only prove to be less efficient, but can actually be detrimental to the properties of the concrete. In precast system, the normal maximum curing temperature has been found to be 160 F [4] [13].

It's very much possible to accelerate the curing process, so that it occurs within a shorter duration. The major methods for accelerating the rate of the curing process include: physical methods, additions of mineral admixtures and chemical admixtures. The physical processes for accelerated curing can be accomplished through simple conduction/convection via the circulation of hot water or oil through formwork, or even through pipes inside the concrete members in the case of hollow elements. Electric resistance heating, low pressure steam curing and high pressure steam have both been employed to accelerate the curing process [13] [14] [15] Examples of mineral admixtures for accelerating curing include: Microsilica, Fly Ash (as Cement replacement) [15], while those of chemical admixtures for accelerating curing are Calcium, High-Range Water Reducers (Super Plasticizers),

2.4. MOVING AND TRANSPORTING A PRECAST COMPONENT:

Any precast component that is to be moved and transported must be adequately reinforced and strong enough to be moved. During the design and construction of precast components, it's very essential to include specified lifting points, which were properly reinforced [16]. Normally, the extent of these reinforcements is hugely determined by what the components are meant for [16]. For instance, a sound-barrier panel, whose function is to resist wind loads, will surely be lightly reinforced. If the panel is placed horizontally during its construction, then it will have to be lifted from the top after the construction. In this case, there will be need for an additional reinforcement that will be aimed at the preventing cracking. For every precast element, the lifting hooks and connections to be used must be strong enough. Such materials should also be placed in the designed position. The essence is to prevent the component from fail. Such accident will lead to cracking or even breakage of the precast materials.

Figure 4: Precast pedestrian bridge deck being lifted in place. Lifting points typically coincide with the designed support positions.

Martin (2013) [16]

Figure 5: Reinforcement around a lifting point in a precast beam

Source: Martin (2013) [16]

CHAPTER 3

PRESTRESSING

As already stated earlier, prestressing is a specialized method that is normally used to overcome the natural but tensional weakness of concrete. It's being widely applied, in the production of bridges, floor, beams etc. The compressive force that balance the tensile of the concrete material are usually provided with the help of the clamping load, through the provision of prestressing tendons. This is quite different from the traditional reinforced concrete, which basically involves the use of steel reinforcement bars, rebars, inside poured concrete. The two main methods of accomplishing preestressing are: pre-tensioning and post-tensioning.

3.1. METHODS OF APPLYING PRESTRESS

Pre-tensioning

In this method of prestressing, the tendons are cautiously positioned within the formwork, before applying the design load or tension. Once the set-up has been tensioned, the concrete mix is then cast and allowed to harden until it achieves sufficient strength to resist the forces, which will be applied to it. This force to be applied is usually above 30 MPa. Once the concrete has hardened, the steel tendons will then be release from the restraints, after which the stress in them are transferred to the concrete by the bond between the two materials. In pretensioning, the commonly used tendons are in the form of small-diameter wires or a combination of smaller wires, otherwise known as strands. It's very important to make sure that the diameters of these materials are small. The essence is to increase the surface area that is available for bonding with the concrete. The best material for increasing the bond is crimped or indented wire [15] [17].

Post-tensioning

In this method of prestressing, the concrete is first allowed to harden before being subjected to the tension. Because of this; the steel tendons cannot be allowed to bond with the concrete, at least not initially. In other to accomplish this, the tendons are positioned in ducts or holes that have been cast in the concrete. Alternatively, the steel tendons can be greased and then sheathed to eradicate any chance of bonding. The same objective can still be accomplished by fixing the tendons to the outside faces of the member [17].

After the concrete has been casted and strengthen to a sufficient strength level, the tendons will be tensioned and then fixed in special fittings cast into the ends of the concrete member. There are specially designed patented fittings and systems that can e used to achieve this objective. Once tension has been obtained, the ducts will then be filled up with cement grouts. The tendons will eventually bond to the concrete once the cement grout set [18].

3.2 MATERIALS AND COMPONENTS

The two most important materials that are needed for the prestressing process are concrete tendons.

Concrete: High-tensile strength proffers lots of opportunities in the construction industry. The efficient and judicious applications of these opportunities can be accomplished through the use of concrete. After 20 days of casting, a typical concrete accrued strengths in excess of 40 MPa. The specific strength that is achieved in any particular case is usually determined by such factors like proportioning, batching, mixing, placing, compacting and curing. Thus, special attention must be paid to these factors, during the production of the concrete. Failure to do so is likely to result in failure of the unit or member being stressed. In some case, the cross-section area of the prestressed members can be so small that it becomes more difficult to place and compact stiff, high strength mixes. Thus, it's very important to always dictate the minimum size of the prestressed member. This can be deduced by measuring the size and spacing of the prestressing ducts and the other reinforcement. The cover to reinforcement can be reduced through the use of high-strength concrete permits. However, the fabrication and fixing of the reduced reinforcement must be done cautiously to maintain the appropriate tolerances [19].

Tendons: Tendons for prestressed concrete are usually manufactured from high tensile steel. Most of the ones used in prestressed concrete are 'low relaxation stress-relieved', while the remaining are 'normal stress-relieved'. Under high stress, tensioned steel held at constant length tends to lose some stress. But Low-relaxation steel has the capability of maintaining higher tensile stress over time than normal steel. This explains why low-relaxation steels are commonly specified in the manufacturing or prestressed concrete structures. Therefore, it's very important for structural engineers to know that, normal relaxation steel cannot be used in place of low-relaxation steel [4] [17].

Every prestressing tendon is expected to have mill certificate that is either issued by the manufacturer or a nominated testing authority. It's also very important to take notes of both the chemical and physical properties of the steel to be used. The steels used in prestressing are normally sensitive to rusting, kinks, notches and heat. The steels should be protected against rusting by storing them under dry environments. Studies have shown that the bond characteristics of tendons are greatly improved by light, hard oxides found on the tendons. Thus, these oxides are both desirable on the tendons and in bonded post-tensioned work.

The smaller diameter of the prestressing steel, in comparism with ordinary reinforced bars, makes the former to be more sensitive to corrosion. Both corrosion and pitting greatly lead to a significant reduction, in the cross-sectional area of the tendon. Consequently, it's very necessary to always check the tendons for early indications of pitting corrosion. Any tendons with the slightest sign of pitting corrosion should be rejected [17]. The tendons should also be checked for notches and kinks, from time to time.

Tendons are also susceptible to excessive heat. This feature is capable of destroying or altering the high-tensile characteristics of the steel. Because of this, any form of welding operations, should not be executed on prestressing tendons. Tendons come in various forms such as wires, strands, or of bars.

Ducts: In prestressing, the commonest method for forming duct is through the use of flexible metal tube. Other methods that can use include: plastic tube, inflatable rubber tube and removable steel former. These ducts are formed in concrete by casting the appropriate tube. The tubes are generally thin, but still have the ability to withstand great stress.

Prior to concreting, the tendons are cut to length and then positioned within the duct. Through this way, the position of the duct will be retained during concreting; thereby avoiding the major problem of trying to thread the tendons through the duct should it be damaged. In situation, where it's not possible to position the tendons in the duct prior to concreting, the position of the duct may still be maintained with the help of a plastic tube, bundle of wires or inflatable rubber duct [17].

It's also possible to make use of straight ducts. The efficient use of this can be boosted with the help of a slightly undersized steel tube. If inflatable rubber formers are to be used, then they should be supported and inflated according to the manufacturer's instructions. Inflatable rubber formers usually require support at 300 mm centres. It's also possible to inflate this class of ducts to pressure of about 200 kPa. Special attention should always be paid to the value of the pressure, especially when the rubber formers are being used in steam-cured concrete. Since temperature is directly proportional to pressure, there is always the risk of increasing the pressure to an unacceptable amount, when rubber formers are used in steam-cured concrete. In other to avoid this, the rubber formers should be deflated in steps. The process must be stopped, immediately the concrete has set and begins to harden. The use of removable steel formers is restricted only to small lengths. This is because of the risks that are normally associated with its usage. Greasing of the tubes to prevent bonding with the concrete can interfere with subsequent bonding of the grout.

Anchorages: These consist of units and components, which make it possible to stress the tendons and then transport the force in the stressed tendons to the concrete member or structure. Anchorages are usually integrated into facilities. The essence is to ensure a successful injection of cement grout into the ducts, thereby protecting the tendons from corrosion and bond them to the concrete.

3.3 CONSTRUCTION

Basically, a prestressing system consists of methods and equipments, which can be used to tension and fix tendons so as to enable them to transport their load to the concrete. Prestressing system comprises of anchorages, the tendons and even the jacks that are used to stretch the tendons. The main steps in the construction procedure are all discuss below.

The Fixing of Ducts and Anchorages: In prestressing construction, fixing and securing the ducts and anchorages within the framework, and preventing them from moving during concreting, is a very important step that should never be overlooked. This process should be executed cautiously, in other to maintain the profile of the elements that are being constructed. This is normally accomplished by tying the duct to supports like, reinforcing steel, chairs, etc. The fixings of both the ducts and anchorages should be sufficiently rigid and close to the centers, so as to prevent incidents of displacement of duct, during the concrete operation. If large duct are being used in the construction process, then the support must be at 3m centres or less (the exact location should be determined by stiffness of the ducts). When small ducts are being used, the support system should be placed about 1 m centres [12] [17].

In addition to displacements likely to be caused by the weight of concrete and the operation of the vibrator, displacement can occur due to flotation of the ducts. This incident can also be avoided by proper fixing of the ducts and anchorages. It's also very important to always monitor the duct, and make sure they are positioned adjacent to the anchorages. Through this way, unintentional angular deviations are avoided.

The main reason for fixing the anchorages is to prevent movement during the process of concreting. If the anchorages are to be attached to the end formwork, then the formwork should be rigid enough to overcome the horizontal forces, which will be exerted on the anchorages, during the concreting operation. Also, the the fixing detail should be such that the ingress of grout at this point is prevented.

Concreting: This is another important phase that must be undertaken cautiously. Any slight defect in the chosen concreting program can jeopardized the stressing operation. It's very necessary for the engineers to inspect the tendons and anchorages, before proceeding with the concreting program. For instance, it's mandatory for both the tendons and anchorages, to be firmly tied at all location, prior to concreting. This helps to eradicate any chances of mortar, escaping into the duct or anchorage device, during the placing and compacting of the concrete.

It's also mandatory to maintain a proper tendon alignment, prior to the concrete placement. The ductwork should generally be handled, cautiously, in order to avoid damaging it. The laborers should avoid stepping on the duct, so as to avoid damaging it. If the ducts are accidentally damaged, then it must be urgently repaired. Failure to do so may prevent the concrete from bonding to the tendons. The repairing processes aren't all that difficult. Small holes can be fixed with waterproof adhesive tapes, while bigger ones can be covered with metal stripe. However, the overlap around the duct must be up to 100 mm and the joints should be sealed by a waterproof adhesive tape. Special attention should be given t grout pipes, air bleed, grout pipes and other end-zones. This should be done to prevent unnecessary voids and maintain uniform compaction across the concrete structure. Once the concreting has been done, the cables (where applicable) should be pull back and forth, in other to make sure they remain free.

Stressing: Hydraulic jacks are used by most stressing system to tensioned steel tendons. Scientists have developed huge hydraulic jacks that are capable of exerting up to 1,800 t members that are being prestressed. These types of hydraulic jack are currently used for the stressing operations in large dams and similar applications. Majority of hydraulic jacks that are commonly used construction sites, are capable of exerting about 300 t (9).

The steel tendons can either be tensioned singly or in groups. In the first case, each of the tendons is individually stressed, whereas in the later case, all the tendons are stressed at the same time. In the single tensioning, small jacks are always used. These small jacks are quite easier to handle, but require more number of operations. On the other hand, the multiple tensioning requires large jack, which are more difficult to handle, but reduces the number of operations needed to bring about the desired effects. The type of jack to use in each case is normally determined by the pressing system in place, as well as the dimension of the tendons.

The timing and sequence of the stressing procedure must follow certain patterns, specified by the designer. However, it's necessary to observe some other rules in course of the stressing procedure. These include [4]:

- Early partial stressing: This is aimed at reducing the chances of shrinkage cracks. It also help to balance the self weight of the slab and to enable formwork to be more economically used.
- The proper stressing sequence: This help to eradicate huge differential stresses in adjacent cables or areas.
- In stage prestressing: This helps to balance the loads that are being applied as the structure progresses.

Grouting: In some countries, the post-tensioned tendons are normally grouted in their ducts. This usually occurs after the stressing operations. The main reasons for grouting are as follows:

- To establish a reliable and sustainable bond, between the concrete member and the stressed tendon
- To protect the tendons in case of any eventuality. For instance, exceeding the ultimate strength of a member can lead to the development of small cracks. Thus, it necessary to surround the tendon with cement-rich grout or concrete, that is both impermeable and compacted.

Generally, grouting should be carried out within seven days of stressing. The period can even be shorter, if the prestressing is done under aggressive environments. The grout used for grouting of prestressing ducts, consists of cement and water. Admixtures, which are capable of improving the properties of the grout, can also be added. Examples of properties to look out for include:

- ability to increase workability;

- ability to reduce the bleeding,
- ability to expand the grout

Materials that are capable of damaging the steel or the grout itself must be avoided. Some examples include: chlorides, nitrates and sulphides.

CHAPTER 4

FUTURE OF PRECAST TECHNOLOGY

Concrete is popularly known as the most widely used building materials in the world. Precast concrete components are providing a very sustainable venue, for expanding the use of concrete. Today, precast technology is being widely applied in both civil and industrial construction. It's just the most suitable choice for our contemporary building technology. Its less energy consumption, eco-friendliness and ability to withstand climatic elements, has made it the sustainable option, which structural engineers, have been looking for ages. Precast technology boost the sustainability of the construction industry by reducing the impact of construction activity on our environment. [20]

The immense benefits of precast technology, means it has a critical role to play in future construction activities. Its environmental-friendliness, affordability and quick to use, are just the features that are needed in the construction industries. In coming years, more and more construction projects will require precast technology. These include: advanced modern factories, large-scale infrastructure, online shopping warehouses, tall office buildings and tall residential buildings. The construction of these buildings will involve high pressures on the construction time-frame, as well as the need to use special facilities. Fortunately, precast technology is very much capable of taking care of these needs. The technology will also continued to be used in the construction of such large scale infrastructures like: longspan bridges, high speed railway lines, subways, and stadiums [21].

CONCLUSION

Precast concrete remain the commonest building materials in the whole world. The numerous advantages associated with its usage, means that this technology will continue to remain popular in the construction industry. However, requirements for sustainability are quite dynamic and are therefore being constantly updated to suit the prevailing need at a particular point in time. Thus, innovation is the most reliable tool that can continue to keep precast concrete structures on the right track. Fortunately, the cost of production is quite low and the required raw materials are readily available. However, there is need for more concentration on the sustainability aspects of precast concrete technology. For instance, even though production of precast concrete is energy efficient, more energy saving strategies should be devised. This can partly be accomplished by using lightweight concrete and embedding heatproof material.

BIBLIOGRAPHY

1. **Allen, EA.** *Fundamentals of building construction materials and methods.* Hoboken, N : John Wiley & Sons, Inc, 2009.

2. **McCormac, J.C.M.** *Design of reinforced concrete.* Hoboken, NJ : John Wiley & Sons, Inc, 2006.

3. **NPCA.** *THE LITTLE BOOK OF CONCRETE: A GUIDE TO THE ONE HUNDRED ADVANTAGES OF PRECAST CONCRETE.* Indianapolis : NPCA 10333 N. Meridian St., Suite 272 Indianapolis, 2006.

4. **Hiroshi Mutsuyoshi, Nguyen Duc Hai and Akio Kasuga.** *Recent technology of prestressed concrete bridges in Japan: IABSE-JSCE Joint .* Dhaka, Bangladesh. : 978-984-33-1893-0 , 2010.

5. **Tang Hui1, Wei Wenlong1 & Cui Xiupeng1** *Preflex Prestressing Technology on Steel Truss Concrete-Composite Bridge with Medium Span..* 1, s.l. : Modern Applied Science: Published by Canadian Center of Science and Education, 2013, Vol. 8. 1913-1844 E-ISSN 1913-1852.

6. **Paper, CEBE Working.** *Learning from Manufacturing: JIT and MRP in Built Environment Education.* s.l. : JIT, 2013.

7. **West, M.** *PRESTRESSED FABRIC FORMWORKS FOR PRECAST CONCRETE PANELS, Director: C.A.S.T. Centre for Architectural Structures and Technology.* s.l. : University of Manitoba Faculty of Architecture, 2012. http://www.umanitoba.ca/cast_building/assets/downloads/PDFS/Fabric_Formwork/Fabric-Formed_Precast_Panels.pdf.

8. **Hardas, M.K.** *Formwork for Precast - An Overview.* 2013. http://www.masterbuilder.co.in/data/edata/Articles/April2013/162.pdf.

9. **Paradigm.** *PRECAST CONCRETE STRUCTURES: Structural Engineering & Geospatial Consultants.* 2013. http://www.paradigm.in/Downloads/5%29%20PRECAST.pdf.

10. **Ghaib, M.A.A. and Gorski, J.** *Mechanical Properties of Concrete Cast in Fabric Formworks.* s.l. : Cement and Concrete Research 31, 2001. pp. 1459-1465.

11. **Vollenweider, B.** *Various Methods of Accelerated Curing for Precast Concrete Applications,.* s.l. : CE 241: Concrete Technology Paper #1, 2004.

12. **Pfeifer, Donald W.***Development of the Concrete Technology for a Precast Prestressed Concrete Segmental Bridge.* s.l. : PCI Journal, 1982. 78-99.

13. **Heritage, Ian, Fouad M. Khalaf, and John G. Wilson.** *Thermal Acceleration of Portland Cement Concretes Using Direct Electronic Curing,.* s.l. : ACI Materials Journal, , 2000. 37-40.

14. **Mehta, P. Kumar and Paulo J.M. Monteiro.** *Concrete, Microstructure, Properties and Materials.* 2001.

15 **French, Catherine, Alireza Mokhtarzadeh, Tess Ahlborn, Roberto Leon..** *High-Strength Concrete Applications to Prestressed Bridge Girders: Construction and Building Materials.* Great Britain : Elsevier Science Ltd, 1998, Vol. 12. 105-113.

16. **Martin, W.J.** *PRECAST ELEMENTS.* 2013. www.nra.co.za/content/Chapter_10.pdf.

17. **CCANZ.** *Guide to Concrete Construction: Prestressing.* 2012.

18. **Institute, Post-Tensioning.** *INTRODUCTION TO POST-TENSIONED CONCRETE.* s.l. : PTI EDC-130 EDUCATION COMMITTEE, 2012.

19. **Michael, Dr Antonis.** *Materials and Prestressing Methods.* s.l. : Department of Civil Engineering Frederick University, 2012.

20. **X, Lu.** *Precast concrete structures in the future: Structural Concrete 15.* 2014.

21. **TNN.** *Precast technology: The latest for quality of construction.* s.l. : http://economictimes.indiatimes.com/precast-technology-the-latest-for-quality-of-construction/articleshow/21163272.cms, 2013. http://economictimes.indiatimes.com/precast-technology-the-latest-for-quality-of-construction/articleshow/21163272.cms.